Be a Science Detective

Victoria Kosara

Contents

You Can Be a Science Detective

Detectives solve mysteries. There are mysteries in nature, too. You can use your thinking skills to solve some of them. You can be a science detective.

Think like a detective as you read this book. You will be thinking like a scientist, too!

How Do Fish Breathe?

When you swim underwater, you have to hold your breath. Fish swim underwater all the time. How can they breathe without ever coming up for air?

Science Secret

When you breathe, your body gets **oxygen** from the air you take in. Fish get oxygen from water as it moves through their **gills.**

Be a Science Detective

Whales swim underwater much of the time. They are not fish, and they do not have gills. How do you think whales get oxygen?

Why Do You Sometimes See Your Breath?

When the weather is warm, you cannot see your breath. On a cold day, you can see your breath when you breathe out. What is the reason?

Science Secret

Your breath has **water vapor** in it. You cannot see water vapor, but it turns into tiny droplets of water when your warm breath goes into the cold air. Then you can see the little cloud that the droplets form.

Be a Science Detective

When water boils in a kettle, it turns into water vapor. The hot water vapor goes into the cooler air. Why do you think you can see a little cloud near the kettle's spout? What **conclusion** can you draw?

What Makes a Shadow?

When you go outside on a sunny day, you can see your shadow on the ground. How does it get there?

Science Secret

Your body gets in the way of the sun's light. You block the way the sun's **rays** are going. Something that blocks the path of light rays makes a shadow.

Be a Science Detective

What happens if your hand blocks the light from a lamp? Why?

What Are Nests For?

Have you ever climbed a tree and seen a bird nest on a branch? Why did a bird build the nest? What is the nest for?

Science Secret

A nest is a safe, soft place for a bird to lay eggs. After the eggs hatch, the nest is a safe place for the baby birds. They stay in the nest until they are old enough to fly away.

Be a Science Detective

Look at these baby squirrels. Why do you think squirrels make nests?

Why Does the Moon Look Small?

When you look up at the moon at night, you can cover the whole moon with your thumb. But the moon is really much, much bigger than your thumb! Why does it look so small?

Science Secret

The moon is far away from Earth. When you see something that is far away, it looks small. If you were on the moon, Earth would look small and the moon would look big.

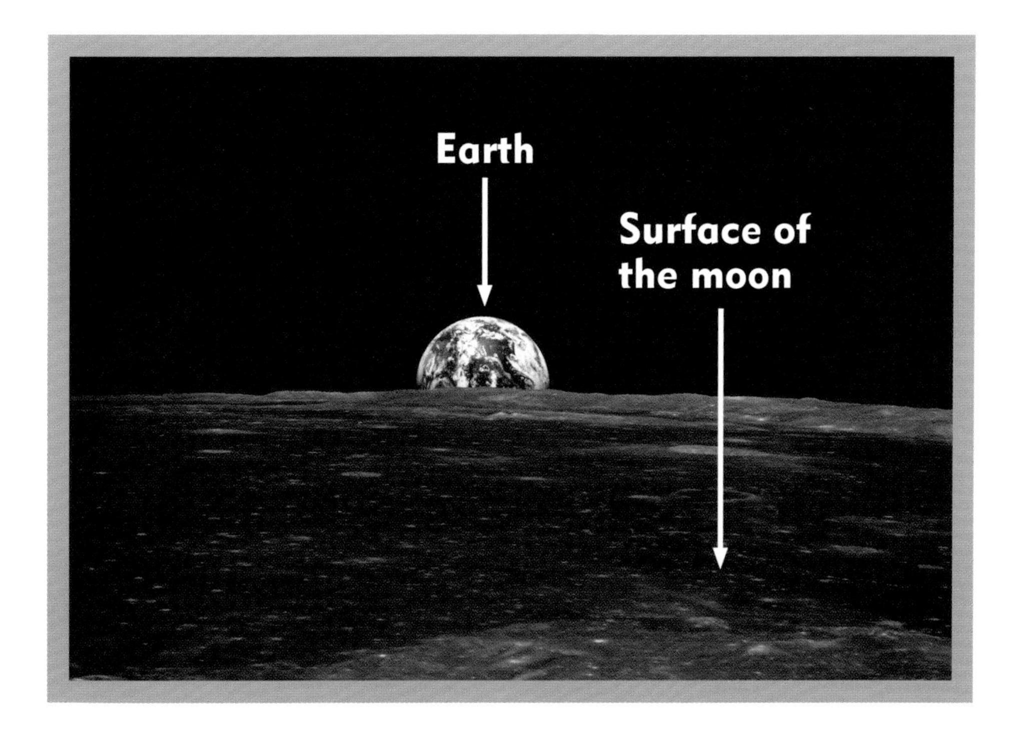

Be a Science Detective

Many things in space cannot be seen at all without a telescope. Why not?

More Science Mysteries

You can find science mysteries everywhere! The more science secrets you know, the more mysteries you can solve.

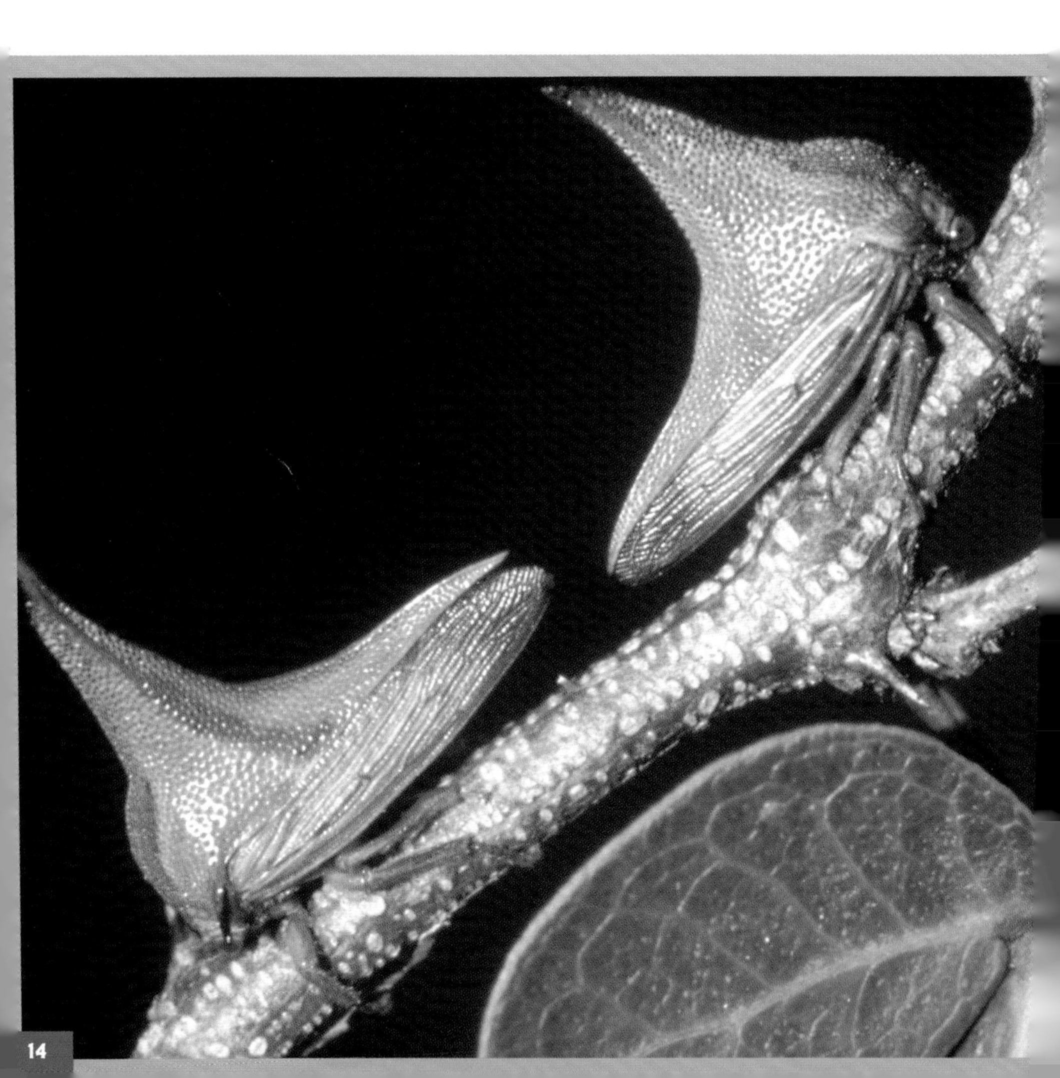

What science
mysteries would
you like to
solve next?

Glossary

conclusion (kun-kloo-zhun): an idea or answer reached by careful thinking about all the facts

gill (GIL): a body part for getting oxygen from water. Fish breathe with their gills.

hatch (HACH): come out of an egg

oxygen (AHK-sih-jun): an important part of Earth's air and water. All animals and plants need oxygen to live.

ray (RAY): a thin beam of light

water vapor (WAW-tur VAY-pur): a form of water that we cannot see. Air contains water vapor.

Index